A SHORT CRITIQUE OF CLIMATE CHANGE

ELGIN HUSHBECK, JR.

Topical Line Drives

Volume 25

Energion Publications

Gonzalez, FL

2017

Unless otherwise noted, all temperature references are in degrees Celsius.

ISBN10: 1-63199-393-3
ISBN13: 978-1-63199-393-0

Energion Publications
P. O. Box 841
Gonzalez, FL 32560

pubs@energion.com
energion.com

ACKNOWLEDGEMENTS

While there are many people that have influenced my opinions on this issue over the four decades I have been considering it, I would like to particularly thank Chris Eyre, a self-styled supporter of the need for immediate action to counter global warming who has challenged me to be better in our discussions, and who gave me insightful critique of an early draft of this book. I also greatly appreciate the extremely valuable editing and comments of Helen Wisniewski. And of course, I must thank my friend and editor Henry Neufeld for his efforts and support.

PREQUEL: THE NOBEL PRIZE

The Nobel Prize, since its establishment in 1895, has become the symbol of singular excellence in broadening the horizons of a field of knowledge. It declares, someone has not only mastered the field, but has made a significant contribution to expanding it. Marie Curie won the prize twice, once in Physics and again in Chemistry. Albert Einstein won the prize for his work in Theoretical Physics and the discovery of photoelectric effect, while Werner Heisenberg won the same prize 12 years later for the creation of Quantum Physics.

To be a Nobel Laureate says something and gives instant status and clout to a man or woman of science. Thus, it is not too surprising in the ongoing debate over Climate Change to see so many of those raising the alarm about the impending danger pointing to their (or the scientist's they are citing) status as a Nobel Laureate.

Rajendra Pachauri, for example, until recently the chairman of the premier Climate Change organization, the Intergovernmental Panel on Climate Change (IPCC), has frequently been described as "a Nobel Laureate" or as a "winner of the Nobel Prize", as in this press release: "Speaking at today's BMW Group sustainability summit in Berlin, Dr. Rajendra Pachauri, Nobel Laureate and Chairman of the UN IPCC..."

Not only do they make the claim for themselves, numerous high government officials, such as U.S. Secretary of State and Prime Ministers, along with government and non-government organizations like UNICEF and the World Wildlife Fund, and of course news outlets like the BBC and New York Times have likewise pointed this out about Pachauri and others working for the IPCC. After all, can you blame them when the stakes in the Climate Change debate are so high?

There are at least two problems with this; the first being that the prize in question is not the Nobel Prize in Physics or even Chemistry, fields that would relate to the science behind Climate Change. The 2007 prize being referred to is the Nobel Peace Prize. It was not won because of the scientific work on Climate Change, but for educating the public on the dangers of Climate Change

2

based on the belief that, if left unchecked, Climate Change would lead to war.

The second problem really came to forefront in late 2012 when Michael Mann, the originator of the iconic Hockey Stick graph so key to Climate Change, and another of the many IPCC members prominently claiming their status as a Nobel Laureate, filed a lawsuit against some journalists and publishers criticizing his scientific work. In the words of his complaint filed with the court he argued they had tried "to discredit consistently validated scientific research through the professional and personal defamation of a Nobel prize recipient."

Yet Pachauri, Mann and the others claiming to be Nobel Laureates based on the 2007 Nobel Peace prize are not, with one exception. The prize was given to the IPCC and Al Gore, so while Al Gore can legitimately make the claim, being part of the IPCC does not make one a Nobel Laureate any more than does being part of any of the many other organizations that have won the prize. Thus, shortly after Mann's complaint was filed with the court, he had to amend the complaint and the IPCC posted on its website, the Peace Prize "was awarded to the IPCC as an organization, and not to any individual associated with the IPCC. Thus it is incorrect to refer to any IPCC official, or scientist who worked on IPCC reports, as a Nobel laureate or Nobel Prize winner."

While ultimately of little importance in the overall debate over Climate Change this does highlight many of the key problems; the appeals to authority, inflated claims, media exaggeration, and how things change so drastically when one looks behind the curtain to see what is really going on.

WHY CLIMATE CHANGE

This book originally started as an article on how agenda often takes precedence over truth and how damaging this can be. I believe claims without examples are largely meaningless and so needed a well-known example. In my opinion, the most glaring current example is the alarmism over Climate Change. Since those taken in by the alarmism would not accept the premise, I started to write about the problems, but soon realized this would need to be more than just one or two; I would need to go deeper to show the theory is at best problematic.

It is important to make clear this remains only a brief overview and I have skipped lots of information. This book is not meant to be a complete in-depth analysis of the pros and cons of the various issues involved in the climate change debate. It is a short critique of the dominant view presented by the popular culture.

I will be presenting a some of the reasons why I question that human caused Climate Change poses a serious danger. If you accept the danger is serious, my goal is not to convert you, but to provide enough information so: 1) you realize there are legitimate reasons to question the orthodox view 2) you become more skeptical of the orthodox position and 3) you are interested enough to investigate this issue more in depth. However, if all you do is come away with a better understanding of the problems that cause many to question the theory, this book will have served a valuable purpose.

CONCERN

Many on both sides of this debate agree, supporters and skeptics alike, if we get the answer to this issue wrong, countless numbers of people will suffer and many will die. Supporters see the damage as more long term with the destruction caused by increased temperatures and rising sea levels. Skeptics see the danger in the near term as the burdens of the proposed remedies leave people worse off, or in poverty and starving. Either way people's livelihood, well-being, and lives are at stake. Thus, if you are concerned about people, you should be concerned about the issue of Climate Change.

While certainly less important, another concern for me is the corruption of science I see taking place in the Climate Change movement, a movement that extends far beyond the scientists. I believe Climate Change will eventually be seen as the largest fraud in the history of science. It will damage the hard-earned respect and trust science has earned for decades, if not longer. I believe this damage is already happening, and the only question is how much damage will be done and how long it will take for science to recover. In their attempt to justify political changes, advocates have effectively moved the theory that humanity faces a serious threat outside the boundaries of science into pseudoscience. In some cases, even into a religion with dogmas that cannot be questioned, but only accepted, with any dissent being punished.

Scientific Certainty?

Why can I so confidently make such statements? To begin with, consider the following key claim by the IPCC, a claim which they have made repeatedly, and is found in the fifth IPCC report finalized in 2014.

> *The evidence for human influence on the climate system has grown since the IPCC Fourth Assessment Report (AR4). It is extremely likely that more than half of the observed increase in global average surface temperature from 1951 to 2010 was caused by the anthropogenic increase in GHG concentrations and other anthropogenic forcings together.*

The Fourth Assessment Report was in 2007, so this sounds impressive as the IPCC defines extremely likely as "over 95%." Yet there is the Hiatus or the Pause, i.e., the fact that warming stopped in the mid-1990's and did not start up again until 2015, and this was likely due to the beginning of an El Nino, not CO_2 emissions. The first question becomes how could they be so confident in 2014 that human actions caused most of the warming, when the planet had not been warming for over 16 years by that point? How could the evidence that we cause the planet to warm grow, when the planet was not warming?

The Hiatus is so problematic, some supporters try to deny it even exists, but many scientists who support Climate Change know it happened and have attempted to explain it in such a way as to be consistent with Climate Change theories. An article in the January

15, 2014 edition of Nature titled, "Climate change: The case of the missing heat" is just one of many trying to solve the problem caused by the Hiatus.

The basic problem is that Climate Change, which is based on computer models of the climate, predicted we would warm, but we did not. Normally this would call the models and thus the theory into question, but that never seems to happen with Climate Change. So, the question becomes; where did the heat go?

There have been numerous suggestions put forth. The Nature article above being just one of many. It looked to the oceans, suggesting, "perhaps" the El Niño of 1997–98 tipped "the equatorial Pacific into a prolonged cold state that has suppressed global temperatures ever since." Perhaps it did, perhaps it didn't. Maybe the explanation is in one of the other theories put forth, or perhaps, maybe, there just wasn't any warming.

As the Hiatus went on and became more uncomfortable (how do you proclaim the danger of warming when there is no warming) other scientists went back over the data and wouldn't you know it, they found there had in fact been warming. This is not surprising as there are a tremendous number of assumptions that must be made in these calculations. In fact, this is one of the problems as things are so complex and there are so many variables and unknowns, all it takes is a few tweaks to a few variables and you get a different answer.

For example, it is well known there are Urban Heat islands, i.e. when people come together in groups they warm up the immediate area. It can get fairly cold (-30 F) in the winter where I live. My car gives me the outside temperature and as I drive through different municipalities I notice the temperature rise as much as 10 degrees for larger towns (-29 to -19 for example) only to fall back down again as I leave. The increase is not as much in smaller towns, as larger towns produce more heat than smaller ones.

Over the last century many cities have grown with the result that thermometers that were once outside the city and its heat, are now inside the city with it warmer temperatures. These thermometers will show an increase, not because the climate has warmed, but simply due to the fact that the city has expanded to surround the thermometer. This artificial increase, while significant to the local area, is insignificant to the global temperature and must be estimated so it can be removed. How big is this? It changes from city to city and even then the effect is not constant. On windy days the difference between country and city will be small, if it exists at all. Considering the warming we are supposedly seeing from Climate Change is a tiny fraction of the warming of the Urban Heat Effect, it does not take much of a change in the estimate to go from warming to cooling.

The Urban Heat Effect is only one of the many factors that must be estimated in order to reach a global temperature. In the mid-1990s satellites were launched that could measure the global temperature. These satellites would not be affected by many of the sources of error affecting terrestrial measurements, so they were heralded at the time by supporters as being able to definitively show the earth was warming. Yet when the satellite data started coming in, no warming has was seen. Climate Change supporters now believe their launch just happened to coincide with the beginning of the Hiatus. For skeptics, this raises questions about the earlier claims of warming.

Some supporters now claim after looking over the data again, we had in fact warmed a little. But had we? Or does the ability to go back and find warming just cast even more doubt on the pre-satellite data? We are left with some scientists accepting the Hiatus and proposing various theories as to where the missing heat went, while others reevaluated the data and claim to have found some warming after all.

A key problem is that oceans along with the atmosphere and clouds all play a huge role in climate, much of which we simply do not yet understand. We have a major phenomenon, the Hiatus, which so far has not been explained and there are many conflicting views. Given the models were all wrong, and the uncertainty about exactly what happened, how does this square with a claim of 95% confidence? Frankly, it amounts to "We are 95% sure humans are the cause of the warming that should be there but for some un-known reason isn't, but rest assured we do have some theories and one of those might explain it."

While it is a bit more complex than this, basically in science the accuracy of an answer or prediction cannot be better than the inputs. I would know instantly there was a problem with your claim, if I give you a tool that could measure length to plus or minus 0.25 inches, and using that tool you claimed to have measured some sticks and came up with a total length of 33.24385489 inches.

When applied to the climate, the bottom line is we do not really know what is going on with the missing heat, or if in fact there was missing heat, or even much about how the oceans really affect the climate. It was only in the year 2000 that the Argo project began deploying a network of 3000 "floats" to measure the temperature and salinity of just the upper 2000 feet of the oceans. This is certainly a great improvement. As the Argo web site points out,

> *Lack of sustained observations of the atmosphere, oceans and land have hindered the development and validation of climate models. An example comes from a recent analysis which concluded that the currents transporting heat northwards in the Atlantic and influencing western European climate had weakened by 30% in the past decade. This result had to be based on just five research measurements spread over 40 years. Was this change part of a trend that might lead to a major change in the Atlantic circulation, or due to natural variability that will reverse in the future, or is it an artifact of the limited observations?*

Still, while Argo is a great improvement, with 139 million square miles of ocean surface, that is only one float for every 46,566 square miles, which is the size of the States of Vermont and New Hampshire combined. A lot can, and does, happen in 46,566 square miles a single sensor simply misses. If this was the visibility we had on land, we would be unaware of all but the biggest weather events, and things like tornados might still be unknown. And this is assuming an even distribution, which there is not. Supporters must argue that the smaller events and processes that are missed are unimportant and can be ignored, but is this really true?

8

We do not know why the Hiatus occurred and there are many theories. "Perhaps" it was the oceans? Perhaps something else. We really know little about the oceans and their role in the climate, let alone the atmosphere and clouds. So how could any scientist claim 95% certainty? Scientifically they cannot. You cannot start from this much uncertainty and end up at 95% certainty. Therefore, this is not a claim based on science, ultimately this is a political statement, and in this we gain another insight into the problem.

CLIMATE CHANGE

I stated earlier Climate Changes is one of the biggest and most glaring examples of the politicization and corruption of science, but this statement will take some unpacking, lest it be thoroughly misconstrued.

Defining politicization, I mean the moving of an issue into the political realm where acceptance or rejection is governed more by political agendas than the evidence. Put another way, it is where policy begins to drive the science, rather than the science driving the policy. The politicization of science surely contributes to its corruption, but the corruption is much deeper and broader and by no means limited to the issue of Climate Change extending into things such as the problem with the peer review system, as recent investigations have shown.

With Climate Change, the situation is so bad it actually becomes somewhat its own defense, as many supporters simply cannot believe things could really be as bad as they are; that Climate Change could really be, not just lacking in evidence, but, as we shall see, based on falsehoods, and in some cases even fraud and deception.

Language is also a major problem, so a few additional clarifications are necessary. Part of this can be seen in the constant change in how the danger is labeled. It started in the 1970's with global cooling and the coming Ice Age. As temperatures warmed instead of cooling, the new threat became global warming. Again, temperatures did not cooperate and instead of increasing they leveled out, so the new threat became Climate Change, and now we may be on the verge of a yet another change to Climate Collapse.

Actually, the commonly used labels mask a lot of agreement. Both supporters and skeptics agree the climate changes. In fact, as we will see shortly, in some respects those who are supposedly "Climate Change Deniers" actually believe it changes much more than do the supporters of Climate Change. Both sides also agree it has grown warmer over the last 100 years. This is not the crux of the disagreement, nor is the idea our actions play at least some role in this warming.

The points in dispute are 1) how much of a role do humans play 2) whether or not the change is anything to worry about 3) is there anything we can do about it? The belief we do play a significant role, and we should be concerned and take action, might better be labeled *Dangerous Anthropogenic Climate Change or DACC*, i.e. human action is threatening the planet.

I would also point out that when I say that DACC is not only false, but in some cases even based on fraud and deception, I am not saying all those who accept DACC are liars or even dishonest. I believe most have simply been taken in. Nor am I saying all scientists who support Climate Change are being deceptive, but this is part of the problem. DACC is not just the creation of scientists, it is an intermixing of scientists, politicians, and the media. Except for a few individuals, from my perspective, it is the politicians and the media that are the major source of the problem.

Finally, most of those skeptical of DACC are not claiming they understand climate change. At its core the argument of the skeptic is we do not yet know enough to make any reliable predictions. Might the planet warm over the next century? Sure, but it could just as easily cool. Ultimately, we do not know enough yet to say, though given the historical record it is probably more likely the current warming trend will continue regardless of any actions we take however painful and destructive those actions might be. This is why the skeptical argument is the much harder argument to make.

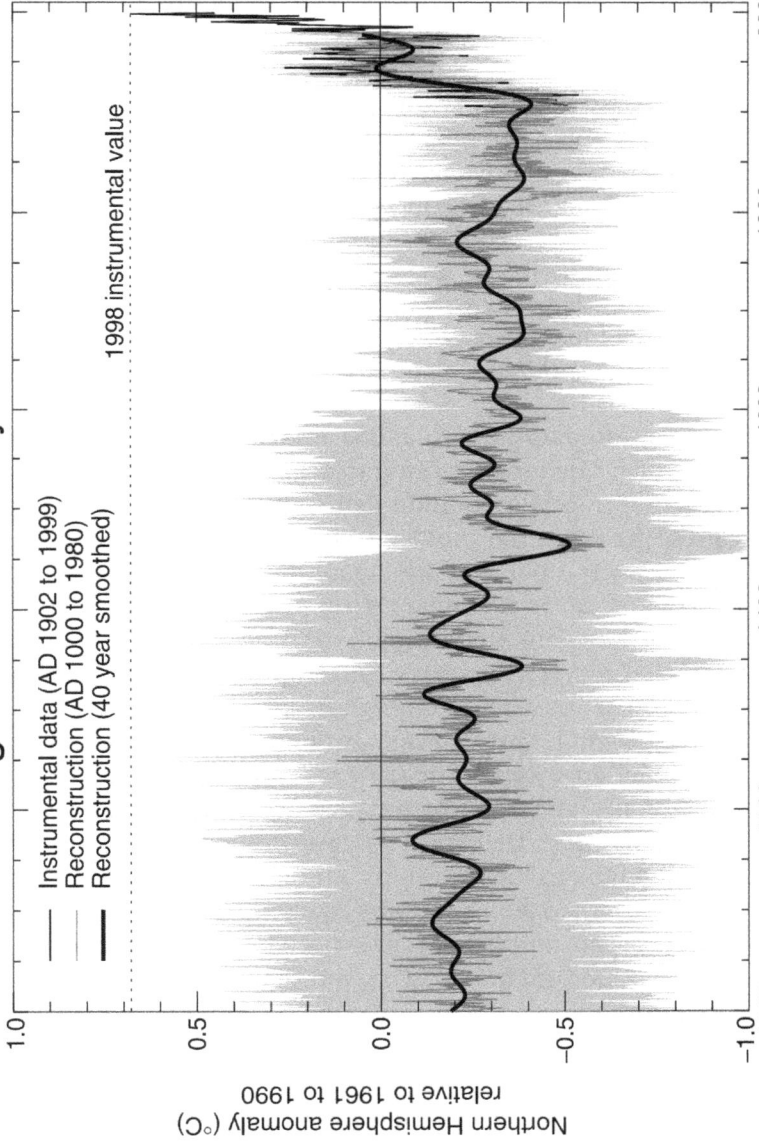

Figure 1 The Hockey Stick

The Hockey Stick

The underlying premise of DACC is best represented by the Hockey Stick graph in Figure 1. This graph was a major component of the 2001 IPCC report and came to represent the danger we face in numerous news reports, documentaries, governmental reports, and school rooms around the world. It purported to show the earth's temperature had been on a slow decline for 1000 years, implying it had always been such. This is the long shaft of the stick. About the time the industrial revolution began, the temperatures started rising at an alarming rate, the blade part of the stick.

The increase purportedly was caused by greenhouse gases released with the burning of fossil fuels and unless stopped, would wreak havoc and could even threaten life on earth. Further "proof" of this was the graph shows temperatures are warmer now than they have been in the last 1000 years, a seemingly impressive claim.

While this could be a clear indication of the danger and its cause, the problem is Hockey Stick graph is false, as is the claim about the current temperature being the warmest. The shaft part of the Hockey Stick was based on trying to determine temperatures by inferring them from a very small number of trees rings. While given the small number of samples this is questionable at best, the real problem is we have several better ways of determining temperature over this period.

These other ways do not show the slow and steady decline found in the Hockey Stick but a much more turbulent picture with what is called the Medieval Warm Period, followed by the Little Ice Age. These swings in temperature are not speculation but part of historical record. We know of Viking settlements with pastures in Greenland that are now frozen year-round, while during the Little Ice Age we have written records and paintings of the much longer and deeper winters, freezing areas that do not freeze now nor during the Medieval Warm Period. The Thames River froze over 23 times during the Little Ice Age; the last time in 1814. At times the river ice was so thick that Frost Fairs were held on it.

The data for the blade part of the Hockey Stick was based not on tree ring data, but on temperature data. Why the switch? Supporters claim the switch was because the temperature records are more accurate, which of course they are. Yet this raises the question, if the tree ring data is not a good indication of temperature in the blade part of the stick, why should we accept it during the shaft part? There is also the additional problem the tree ring data continued to show a decline.

While it took six years and the intervention of the U.S. Congress and the editors of Nature, other researchers were eventually able to get the data and methodologies behind the Hockey Stick graph. The methodology used was extremely prone to producing hockey stick like graphs, at times producing them even from random sets of data. They also found the key data was tree ring data from the bristlecone pine. In fact, this particular data was so central, even given the method's tendency to produce hockey stick shaped graphs, if the bristlecone pine data was removed, the Hockey Stick collapsed.

Making this even more problematic is that bristlecone pine data is known not to be a good source for even regional temperature data reconstructions much less global. Finally, while statistical test scores supposedly showed the Hockey Stick was statistically significant, attempt to reproduce this on the shaft part of the graph failed. When asked, one of the authors later stated they only ran these tests on the blade part, i.e., the part based on temperature records, and thus the part least in question.

A MATTER OF PERSPECTIVE

With the crumbling of the Hockey Stick, so too crumbles a key support for DACC. The claim that temperatures were in a slow decline until humans started burning fossil fuels on a large scale was based on a questionable switch from tree ring data to temperature data, and even then did not match the historical record. However, there is an additional problem as DACC strongly depends on the time frame chosen for comparison.

The recent increase looks alarming if, as it often is, the temperature record is viewed over the last 100 years or less. It is much less so, if viewed over the last 1000 years, with its high in the Medieval Warm Period dropping to a low during the Little Ice Age, with a partial rebound to today. There is no reason for concern at all, if viewed over the last 10,000 years. Figure 2 shows the temperature based on Ice Core data, and this data is not at all unique.

Figure 2 - Relative Temperature from GISP2 Ice Core Data

(Ice cores are cylinders of ice drilled out of a sheet of ice such as a glacier. Studying the ice at various points along the core can yield information about the conditions when the ice was formed.)

This chart shows there have been over 20 similar increases to what we are now experiencing, many much larger. If you are a supporter of DACC, my guess is, you may not have seen this data before, as it is pretty hard to line this up with the claims of DACC. Granted this is not the last word in temperature data, but it does line up far better with the historical record than does the Hockey Stick. Note the increase from about 800 to a little past 1000 is much greater than the recent increase. It was around 1000, in which Greenland was warm enough to get its name. This peak, known as the Medieval Warm period along with the Little Age leading up to the most recent increase are missing from the Hockey Stick.

On the other hand, one can just imagine the panic if supporters of DACC had lived a little over 6000 years ago and faced the change seen there. The point is that the climate is always changing and the increase we are seeing now is well within norm and actually pretty small by comparison. In fact, according to other findings, the last few tens of thousands of years seem to have been marked by relatively mild swings when compared with the even more stark changes found in other periods of earth's history. In short, the climate was changing, and to a much greater extent, long before humanity could have had any influence.

This is hardly the only evidence that has been distorted, or at least not really put into context. The main concern is CO_2 which all agree is a greenhouse gas we are putting into the atmosphere with the burning of fossil fuels. Supporters of DACC frequently label CO_2 a pollutant, leading many to think it is toxic like other pollutants. Yet it is a necessary component of life. Get rid of CO_2 and life on earth ceases. In fact, owners of greenhouses will purchase CO_2 generators because the current atmospheric levels are too low for optimum growth. So CO_2 is in the strange category of a necessary pollutant.

While burning fossil fuel does result in CO_2 (and water vapor with its even greater warming effects), is this really a concern? CO_2 is only 0.04% of the atmosphere and natural sources also release CO_2, in fact much more than we do. While supporters talk about the seemingly large 70 million tons of CO_2 human activity add each day, they normally skip over the part about this being about 0.0012% of the atmosphere. While the current warming trend started at roughly the time of the industrial revolution, it was well into the 20th century before CO_2 emissions reached a level that could have had any effect. What started the current trend and kept it going for over a 100 years?

In addition, plants need CO_2 the way we need oxygen. Doubling CO_2 from current levels produces yields from 28% to 150% higher depending on the crop. Other plants also grow better at higher levels of CO_2. Increased CO_2 levels really would make a greener planet.

This link between plants and animal life form part of an extremely complex system, what in engineering would be called a control loop. Like the thermostat in your home, a control loop uses a feedback mechanism (for the thermostat it is temperature) to control something (the furnace or air conditioner). As more CO_2 is produced plants grow better producing both more food and more oxygen needed by animals. If these were the only factors, the system would spin out of control. The very fact that life on earth has existed as long as it has, surviving things like massive volcanic eruptions and meteorite impacts that have had far more effect on the climate than anything humanity has done, shows the climate system must be both very complex and robust.

We are only beginning to understand how parts of this system work. There are other free back mechanisms that pull CO_2 back out of the atmosphere, and a lot of these we simply do not yet understand. Nor is it likely we will reach any sort of tipping point. We know that at times in the past there have been much greater quantities of CO_2 in the atmosphere, from ice cores and other records, in fact many times more than now.

Many studies have shown temperature is less sensitive to changes in CO_2 than DACC the models assume. This is undoubtedly one of the reasons predictions made by climate models have consistently and significantly overestimated the increase in temperature. Again, the Hiatus is a problem as anthropogenic CO_2 emissions continued to increase during this period while temperatures did not. So while CO_2 is a greenhouse gas, the evidence is the climate on earth is far more complex than simply increased CO_2 leads to increased temperature. In fact, the historical record shows the reverse, warming normally precedes increases in CO_2 rather than being caused by it.

Perhaps the best case DACC has is with concern over the increase in sea levels, but again perspective is often left out. Yes, the oceans are rising. In fact, they have risen about 130 meters over the last 12,000 years, or since the last major Ice Age, most occurring in the first few thousand years as the rate of increase slowed significantly a little over 7000 years ago. Since then it has only risen about 4 meters. Based on evidence from fish pens dug by the Romans, it looks like it has risen about 0.13 meters in the last 2000 years.

Yet, not only do sea levels rise and fall over time, so do land masses. There are, for example, ancient port cities that are now miles from the coast while others are underwater. Thus in the study with the Roman fish pens the actual overall change was 1.35 meters with oceans rising 0.13 while the land sank 1.22.

The overall change of last few thousand years looks very small and are almost level when compared to the 130 meters since the last major Ice Age. Thus, the IPCC can claim, "Sea level indicators suggest that global sea level did not change significantly from then until the late 19th century." Since 1900 we have risen .2 meters. This is just the Hockey Stick in another form. While it is true that sea levels 2000 years ago were very close to what they were in 1900, that is sort of like comparing the temperature in April to the previous October and saying not much changed since October.

Again, we are currently coming out of the Little Ice Age, one of the effects of which was to freeze parts of Greenland, among other places, lowering sea levels and leaving many once thriving medieval port cities far from the sea. They are still far from the sea. If we are entering another warming trend like the Roman Warm Period or the Medieval Warm period we should expect the oceans to be rising again, just as they have in the past.

Some studies show the rate to be increasing; others show it is decreasing, while a study of sea levels on the Tarawa atoll from 1993 to 2011 shows no change. It is also interesting to note, 90% of the sea level "rise" of 1.35 meter in the study of Roman fish pens mentioned above was from sinking of the land. As in all the previous times, while oceans are rising the cause is far more likely to be natural than man-made.

Accurate?

A key question in this debate is whether or not the warming is really something we should be concerned about. Part of the problem when assessing this is supporters really have two drastically different ways of talking about the risk. One is more data focused, the other is quite frankly apocalyptic, i.e., basically the end of the world stuff. The apocalyptic claims are the easiest to deal with since they have so often been drastically wrong. One can pretty much pick their disaster: drought, famine, starvation, hurricanes, tornados, the loss of the ice caps, winters without snow, destruction of the Great Barrier Reef, even war and terrorism all have been linked to DACC. All have failed.

The earliest predictions failed because in the 1960s and 70s global cooling was the problem, cooling that was to spark the next ice age. In fact, the global temperature was supposed to have dropped 4 degrees by 1990 and 11 degrees by the turn of the century. The problem was, at about the time these predictions were being made, the short-term cooling trend which had started following the highs of the 1930s came to an end and we entered the warming trend which lasted from 1970s until the mid-1990s. By the 1980s the warming trend had made the claims of global cooling untenable and thus began the concern about global warming.

Soon warming began to be linked to greenhouse gases emitted by fossil fuels. In 1986 James Hansen predicted the world would warm 2 degrees by 2008 leading to a long list of bad effects, everything from windows being broken by high winds to an increase in crime due to the heat. By 1989, the UN Environment Programme (UNEP) Director was predicting warming would lead to rising sea levels which would wipe entire nations off the map. Of course, none of this happened.

In 1990 others predicted the greenhouse effect would be "desolating the heartlands of North America and Eurasia with horrific drought, causing crop failures and food riots." By 1996 the Platte River would dry up and we would see another dust bowl, as Americans sought refuge by fleeing to Mexico.

In 2000, a scientist at East Anglia predicted within a few years, snow would become rare in Britain. Instead, since then Britain has had several record snow falls. In 2001, the IPCC predicted wheat and rice production would be adversely affected, instead there have been record harvests. In 2003 an Australian scientist predicted a drought in the Murray Darling basin of Australia would deepen, instead the rains returned and the drought ended. In 2005 another Australian scientist predicted that a then current drought was permanent and cities only had a two year supply of water left. Instead, again the rains returned, there was flooding, and since then the reservoirs have filled.

In 2006, Al Gore's film, *An Inconvenient Truth,* featured a claim sea level rise had already forced citizens of the Pacific nations to evacuate to New Zealand. This simply had not happened and instead a study had showed 86% of the islands had in fact grown in recent years. In 2007, it was predicted that arctic summers would be ice free by 2013, instead while 2012 was a low year for ice, 2013 saw a 50% increase in ice, leading some to simply push their prediction out to 2016.

This is but a small sampling of the numerous failed predictions. In addition, all the predictions since 1996-7 took place during the Hiatus, when there was no increase of warming. Yet undeterred by the lack of warming and repeated failure, supporters continued to make new predictions, which have subsequently failed. Why should we believe their predictions of the distant future, given the consistent track record of recent failure?

HARMFUL?

Historically, the IPCC has consistently overestimated warming and of course missed the Hiatus completely. Perhaps this is why the most recent IPCC report in 2014 reduced the projected warming. Now they predict earth will only warm about 1.5-2 °C over this century. This would be warmer, if it happened, but would it be a disaster? Hardly. Based on the temperature data in Figure 2, this would put us a little above the Medieval Warm Period but still below earlier warm periods in human history.

As for whether it would be a disaster, or even bad, depends on a number of factors. This of course would be a major problem if the warming does occur and it caused the sea levels to go up, flooding where you lived. Just as there has been with the natural changes throughout history, with any such change there would be both pros and cons. As a Viking living in Greenland when the Little Ice Age happened, you would have definitely been adversely affected. Same for medieval merchants in port cities left high and dry from the lowering sea level. On the other hand, these changes did not occur overnight. There was time to adapt.

On the whole, will we be better or worse off? As the planet warms in some areas it may undoubtedly become more difficult to farm crops. On the other hand, a quick look at a map will reveal that a very large part of the earth's land masses are high up in the northern hemisphere, vast frozen plains in Canada and Russia, places with lots of land but very few people because it is just too cold. Canada and Russia together make up nearly 20% of the worlds habitable land mass. Warming would help these places to become more habitable. Perhaps even the frozen Viking settlements in Greenland would become habitable again.

Another consideration is people prefer warm over cold. In the United States for example, there has been a fairly steady migration of people away from the colder northern states to the warmer southern states as people seek their own personal form of global warming. While many in the winter express a strong desire for warmer days of spring and summer, far fewer people desire the cold, unless perhaps they are waiting for the ski slopes to open.

There is a more serious side to the warm versus cold issue. A study by the National Center for Health Statistics reports, in the United States twice as many people die each year from weather related causes due to cold than to heat. As a study in the UK estimated the difference between summer and winter deaths was over 31,000 people for the winter of 2012/13. Thus, if the earth does warm, fewer people will die each winter.

Given the desire of people to live in warmer climates, the increase in habitable land, a greener planet, and the savings of lives, the real danger is not to be found in the consistently false predictions of doom and gloom from those concerned with warming, but the fact there is another group of scientists who argue it is very possible this century will be colder than the last one.

COST-BENEFIT

Of course, all of these numerous dire warnings and predictions have a purpose: to persuade people to support a large array of new taxes, programs and restrictions all aimed at reducing CO_2 emissions. While the need is repeatedly stressed the results are often skipped. One problem is there is no worldwide commitment. In dealing with individual countries like the United States, or groups like Europe, the benefit from the proposed changes are so small as to basically be non-existent. Even if the United States reduced it CO_2 emissions to zero, the resulting reduction in temperature over this century would only be about 0.15C which is statistically insignificant, and this is assuming IPCC estimates are correct. On the other hand, the costs of such an effort, both in economic and human terms would be devastating.

To have any meaningful effect the effort must be both worldwide and mandatory, yet so far this has proven impossible to achieve. In many ways, the recent agreement in Paris, despite all the fanfare, was a huge step backwards as it moved away from legally binding commitments, back to contributions without legal force.

The reluctance on the part of nations to commit to reductions shouldn't be surprising. For first world countries such as the United States, the economic effects would be nearly a 5% reduction in GDP and this is the IPCC's optimistic estimates that depend on creating technologies that currently do not exist. Even the small efforts already undertaken have had a negative impact with no discernible benefit.

In addition to reducing the standard of living, increasing unemployment and other economic hardships, the attempts to move to so called green energy have expanded our understanding of poverty to include a new classification: Energy Poverty. Germany, for example, made a major push for renewable energy, only to see their energy bills skyrocket, leaving many struggling to pay them, and 1 in 6 households in Energy Poverty. Some Germans now have to decide between lights and warm meals, while others have to forgo electricity all together as a luxury they can no longer afford.

As bad as a reduced standard of living and energy poverty are for those living in first world countries, this is nothing compared to third world countries where poverty, energy and otherwise, are a way of life for over a billion people. Most of the things making our lives better, safer, and longer require energy. The main difference between the prosperity of the last 200 years and the previous thousands of years of human history is the availability of relatively cheap energy sources. While some might cite other factors such as mass production, or increased production of food, all of those are based on energy sources. It takes energy to power the factories and tractors, along with the trains and trucks needed to transport what they produce.

Thus, is it any wonder developing countries are reluctant to commit to any mandatory carbon limitations? Poverty, including energy poverty, is already killing people. Restricting cheap sources of energy does not just limit the ability of people to light their homes at night it also restricts the availability of refrigeration to preserve food, and electricity to power medical equipment in hospitals and clinics. Yet, some still block new power plants, basically condemning people to die for the sake of a theory that so far has failed.

CONSENSUS?

Usually long before this point, there is an effort to avoid evidence such as this with appeals to authority and consensus. What about the 97% of scientists who support Climate Change? Unfortunately, like the Hockey Stick these claims are also lacking support and hold up only as long as you do not ask too many questions.

Concerning the specific and often cited 97% claim, that comes from a survey sent to 10,257 earth scientists. Only 3,146 responded. As a self-selecting survey we already run into serious problems. From this 77 scientists were selected to represent the entire group and it was out of these 77 scientists that 75 scientists, or 97%, were found to be the consensus. While 97% of the selected sample, this is less than 1% of the 10,257 earth scientists that were sent the survey. This would not necessarily be a problem, but this was not a scientific survey with a random sample.

Frankly, there is a lot of confusion over consensus, the biggest question being, who should be considered? Should Engineers be included or just scientists? Should all scientists be considered, or just those directly working on Climate Change? A leading alternative theory to DACC argues that climate change is driven by the Sun and various natural cycles; should Astrophysicists be included?

Another source of confusion is just what is the consensus on? Is it just warming? Is it that human activity is contributing to the warming? Or is it that humans are the major cause of warming and that the warming is dangerous, i.e., DACC? Again, even many skeptics would agree with the first two. Depending on whom you include or exclude, and what you ask, you can get very different answers to the consensus question.

Frankly, most scientific climate research is directed at a wide range of questions that do not directly touch on DACC. A review of the abstracts of nearly 12,000 climate related scientific papers published over the 21 years leading up to 2011 found that only 0.5% explicitly state human actions were the majority cause of global warming. As for views on Climate Change, a study in 2013 found that only 36% of professional engineers and geoscientists fell into the group that believed "Climate Change is happening, that it is not a normal cycle of nature, and humans are the main or central cause." On the other hand the second largest group, "believe that changes to the climate are natural, normal cycles of the Earth… Humans are too insignificant to have an impact on nature." The rest though, the cause was either unknown (10%) or unlikely to have a significant impact on human life (17%). Other surveys of Meteorologist show similar skepticism.

In the realm of science there is, or at least should be, a debate over the meaning of the evidence, with some scientists on both sides. Science does not advance based on consensus, to the contrary the history of science is full of examples of consensus holding back and even trying to suppress alternative theories which ultimately came to be accepted. Even the law of gravity is not grounded in consensus, but in the vast amount evidence that underpins it. Yet it, like all theories stands ready to be modified or even overturned should evidence emerge that conflict with the theory.

Whatever your view of consensus, the fact remains there are thousands of scientists working in the field who either question or outright disagree with the theory of DACC. They include prominent scientists working at major institutions and universities, including Nobel Prize winners (that is real Nobel Prize winners).

In addition, as mentioned above there are in fact alternative theories put forth by respected, but less publicized scientists.

Science works best, not as a committee, but as scientists examining the evidence, forming a hypothesis, constructing experiments to test them, and then challenging other scientists to review, duplicate, or falsify their work. Science is an evidence based endeavor grounded in testing and verification, not a popularity contest.

Instead of consensus, we have supporters of the DACC in key locations dominating the media. How did the Hockey Stick get such a prominent role in the IPCC Report, a role which allowed it to become so important, if it was so bad and conflicted with what was known? Dr. Rob Van Dorland another of the IPCC reports lead authors stated in 2005, "The IPCC made a mistake by only including Mann's reconstructions and not those of other researchers." How did all the conflicting evidence get ignored? Could it be the lead author of that particular section just happened to be Mann, the originator of the Hockey stick?

Science?

Science is built on tests and verification. A hypothesis must be testable, to be considered scientific. According to Karl Popper, a leading philosopher of science in the 20th century, when distinguishing between science and pseudo-science, "one can sum up all this by saying that the criterion of the scientific status of a theory is its falsifiability, or refutability, or testability."

This is a huge problem for DACC, as it is based on climate models whose predictions have consistently failed. Scientists rarely predict just a number, when trying to model something like global temperature, rather the model produces a high and low range of values centered round the predicted value. This range effectively becomes the boundaries of the model. Yet actual temperatures consistently fall close to the lower boundaries of Climate Change models (i.e., less warming than predicted) and in fact are often lower than even the lowest boundary predicted by the model. In short, the models fail. Yet we are still asked, not only to believe, but to act upon the predictions 50 to 100 years in the future, predictions that are for all practical purposes untestable in advance.

In short, DACC is based on false claims of abnormal temperature increases, misleading claims about the effects of CO_2 in the earth's atmosphere and rising sea levels, grounded in computer climate models that are consistently wrong, and defended with false claims of a consensus. We are in effect being told to forget the tests and just act as if they were true anyway. This is not science, but pseudoscience.

True science is based on testing and verification. Yet leading scientific supporters of DACC ask we accept their conclusions while denying access to the data upon which their conclusion are based, making it essentially impossible to test them. As Phil Jones, one such supporter put it, "We have 25 or so years invested in the work. Why should I make the data available to you, when your aim is to try and find something wrong with it?" Could it be falsifiability, the falsifiability that is at the core of real science? Note how Jones' statement does not reflect a dispassionate search for the truth, but the desire to protect a particular point of view, to keep that view from being tested lest a problem be found that would invalidate it.

CORRUPTION?

This desire to protect an agenda can also be seen when Jones wrote to his colleague Michael Mann the developer of the Hockey Stick, "The two MMs [*McIntyre and McKitrick – researchers that had debunked the Hockey Stick*] have been after the CRU station data for years. If they ever hear there is a Freedom of Information Act now in the UK, I think I'll delete the file rather than send to anyone." Is it any wonder the CRU later reported the data had been "inadvertently deleted?"

A report in early 2017 revealed how the National Oceanic and Atmospheric Administration (NOAA) "rushed to publish a landmark paper that exaggerated global warming and was timed to influence the historic Paris Agreement on climate change." Another report revealed how NOAA altered raw temperature data to create a warming trend.

This desire to protect an agenda can also be seen in the infamous Climategate emails to use "Mike's trick" to "hide the decline." The "trick" was to switch from tree ring data to temperature data for the Hockey Stick. The "decline" was the fact the tree ring data continued to decline into the 20th century. As mentioned earlier, supporters argue that was not really a trick but a mathematical technique, as the temperature data was more accurate. Yet, if the tree ring data was not accurate enough to use in the 20th century, would this not indicate it is was not accurate for earlier centuries as well? More importantly why switch? Why not just show both sets of data?

As Dr. Judith Curry wrote concerning the Hockey Stick diagrams, they "are misleading. I was misled… It is obvious that there has been deletion of adverse data." She summed up the situation as, "this whole issue is a big problem for the science and has been an enormous black eye for the credibility of the IPCC and climate science."

There are the other emails that showed attempts to suppress studies that did not support DACC, the deletion of emails regarding their work and even attempts to have a "troublesome editor" replaced because he considered publishing scientific articles that conflicted with DACC's claims. But we are not supposed to worry about any of these Climategate emails as they were reviewed by independent bodies and it was shown nothing was wrong. We were effectively told, Move along, Move along, these are not the droids you are looking for.

Still despite these assurances, the distortions of climate science became so great they led to the resignation of Harold Lewis from the American Physical Society, in protest. His resignation letter should be read in full as it summarizes much of the problem, but here are some excerpts:

> When I first joined the American Physical Society sixty-seven years ago it was much smaller, much gentler… The global warming scam… is the greatest and most successful pseudoscientific fraud I have

26

*seen in my long life as a physicist. Anyone who has
the faintest doubt that this is so should force himself
to read the ClimateGate documents… I don't believe
that any real physicist, nay scientist, can read that stuff
without revulsion. I would almost make that revulsion
a definition of the word scientist… It was a fraud on a
scale I have never seen, and I lack the words to describe
its enormity. Effect on the APS [American Physical So-
ciety] position: none. None at all. This is not science;
other forces are at work.*

Lewis goes on to describe how he and others collected the necessary 200+ signatures to bring the issue before the Council thinking "open discussion" was "in the best traditions of physics." But, "To our amazement, Constitution be damned, you declined to accept our petition." Rather the council's position was developed in a "secret and stacked committee." Thus, Lewis felt he had no option but to resign.

Lewis is not the only scientist to express concern about how DACC is not only based on a very selective reading of the evidence, but it often distorts the evidence while its supporter seek to suppress competing views.

A major concern with all of this is not just the attempted suppression of contrary views, but the attempts to force adherence. Instead of open inquiry, there is the attempt to enforce conformity. Along those lines the words of Judith Curry are very troubling,

*With regards to climate science, IMO the key issue
regarding academic freedom is this: no scientist should
have to fall on their sword to follow the science where
they see it leading or to challenge the consensus. I've
fallen on my dagger (not the full sword), in that my
challenge to the consensus has precluded any further
professional recognition and a career as a university
administrator. That said, I have tenure, and am senior
enough to be able retire if things genuinely were to get
awful for me. I am very very worried about younger
scientists, and I hear from a number of them that have
these concerns.*

IRRATIONAL?

Disagree with the party line of DACC, and its supporters will go after you. The most common label is "Climate Change Deniers," which is often used implicitly and at times even explicitly as an attempt to link those who disagree to Holocaust deniers (e.g., the headline: "Global Warming Deniers Are The Same As Holocaust Deniers, Except Maybe Way Worse"). Other terms include, "Anti-Science," "Anti-reality," "Anti-intellectuals," "Ignorant," "Know nothings" and of course the most popular stooges of the dreaded "Big Oil."

The common link in all of this is an attempt to dehumanize and dismiss any opposition so as to exclude their evidence from consideration. Truly ironic is while supporters implicitly cast themselves in the role as rational defenders of science and reason, their defense uses tactics that are, at their very core, irrational as they are based on the logical fallacy of ad hominem attack. It would be amusing if it were not so serious.

Failing to get support for their policies, supporters are ratcheting up the rhetoric. Some even call for those who question DACC to be arrested as killers. Recently, a coalition of State Attorneys General was formed to go after those who would disagree with their Climate Change agenda. In short, disagree and we will haul you into court and make your life miserable. In the future perhaps worse. Whatever label you want to put on this, it is not the free and open debate where claims must be testable and repeatable. Rather it is believe and do what we say or else.

BEYOND SCIENCE

Now it is important to note, while this attitude is found among some scientists, for the most part much of these last few problems occur once we have moved outside of the realm of science, and into the realm of politics and the media. It was, for example, the L.A. Times that issued a policy that it would no longer publish letters in disagreement with Climate Change. It is politicians who are trying to punish those who do not accept their political agenda by dragging them into court. Debate is so much easier when the other side cannot, or is too afraid, to speak.

While the suppression of dissenting voices comes mainly from politicians and the media, the effect is the same. A real problem is that, by its nature, suppression is often hard to demonstrate. You cannot hear the voices of those too afraid to speak. You cannot see the interviews that did not occur, or read the articles and reports that were not published. Still, some do try.

Bob Carter was interviewed on the BBC, questioning the latest IPCC report. This sparked complaints from a number of IPCC supporters such as John Aston, who had until recently been the Special Representative for Climate Change at the UK Foreign and Commonwealth Office. Aston wrote that to allow Carter to appear was "a serious lapse if not a betrayal of the editorial professionalism on which the BBC's reputation has been built over generations." Some of Carter's claims were that "Climate has always changed and it always will — there is nothing unusual about the modern magnitudes or the rates of change of temperature, of ice volume, of sea level or of extreme weather events," all claims which are consistent with the evidence but which are inconsistent with the more extreme claims of DACC supporters.

Other news outlets have been forced to apologize for letting even a hint of skepticism appear. Roger Pielke, not a climate skeptic, wrote an article, "Disasters Cost More Than Ever — But Not Because of Climate Change." He supported this by citing the IPCC's own conclusion there is no evidence of an increase in extreme weather. Instead the increased costs were due to the growing wealth resulting in more expensive buildings and homes. Even that was too much for the no-dissent-at-all crowd and the resulting on-slaughter of criticism against the website that published the article forced its editor to apologize. While hopefully not supported by any of the major players, some DACC supporters have organized teams they call "crushers" to suppress anything that runs counter to their agenda.

POLITICS

It is the politicians and those in the media who serve as the primary source of the public's knowledge, along with some supporters, who are applying the pressure and demanding conformity. It is not the scientists predicting small increases in temperature and sea level, but politicians, the media, activists and actors making apocalyptic claims about the end of the world as we know it that are the real problem.

This is further exacerbated by blurring the line between science and politics. The major source of information for DACC is the IPCC or Intergovernmental Panel on Climate Change. While it deals with science, this organization is ultimately a political organization not a scientific one. It was not created out of an objective search for truth, but on the assumption humans were causing dangerous changes to the climate, and "to provide policymakers with regular assessments of the scientific basis of Climate Change, its impacts and future risks, and options for adaptation and mitigation." For the IPCC, the key part of the debate was over before the IPCC was founded and their job was to look into how bad the problem is and what to do about it. As Rajendra Pachauri, until 2016 the head of the IPCC put it,

> Let's face it, we are an intergovernmental body and our strength and acceptability of what we produce is largely because we are owned by governments. If that was not the case, then we would be like any other scientific body that maybe producing first-rate reports but don't see the light of the day because they don't matter in policy-making. Now clearly, if it's an inter-governmental body and we want governments' ownership of what we produce, obviously they will give us guidance of what direction to follow, what are the questions they want answered.

Even worse are the large NGOs like Greenpeace, World Wildlife Fund and a vast myriad of other environmental groups for whom industrialization is an inherent threat that must be stopped.

It is from these later groups that a lot of the really outrageous claims come. At times, this results in humorous news reports, such as the ship with global warming researchers, activists and celebrities who believed the prediction that ice at the poles was disappearing. They got stuck in the ice that was not supposed to be there, along with one of the ice breakers that came to rescue them. (Others who did not follow the IPCC's methods had predicted higher sea ice for that season).

Most of the time, it is just the seemingly endless dire predictions that the world will end if we do not act, which fail and are quickly forgotten only to be replaced with new dire predictions. Thus, it is from this group, rather than the scientists we have been getting most of the world-will-end-in-10-years predictions for the last 60 years. Unfortunately, it is also from these latter groups most people get their knowledge about DACC.

WHY?

Frequently, supporters I talk with have trouble accepting the situation could actually be this bad. I have been asked, "Why would they do this?" Note, while it is irrational to address the issue of motives to show someone is wrong, it is not irrational to answer the question of why after the evidence is considered and a conclusion reached. Up to this point, I have put forth evidence, and have avoided questions of motives. My rejection of DACC is not based on motives, but on evidence. Yet, supporters often ask why would they do this?

Embedded in this question is often an ironic twist. While supporters often have no problem seeing nefarious motives in "Big Oil," they seem incapable of see anything but the purest motive from their side. I would submit this is, in and of itself, a problem. People are people, and there are good and bad people everywhere as there are good people and bad on both sides of this issue.

While it is routine to reject those who disagree as "not a scientist," or "not a climate scientist," or "not the right kind of climate scientist," and therefore incapable of having a valid opinion, those who accept DACC are honored regardless of their background. Thus Al Gore gets a Nobel Peace Prize.

What are the possible options for bias? For the media it is clear. They like the sensational. What is more sensational then the end of the world as we know it, particularly when the person making the argument is a celebrity? As for the politicians, there is the desire for power and control. Regardless of how the climate debate has evolved over the last 60 years whether the problem was cooling, warming, or just change, the answer is always the same, bigger and more centralized government with more control. Finally there is a desire to be important, to be involved in something that matters. What could be more important than saving the world?

There is the old adage of follow the money. Supporters of DACC frequently complain about "Big Oil," and how its money is what is really behind those who disagree with them. One such study showed nearly $560 million was spent over seven years in opposition to DACC and it was claimed this money bought a lot of opposition.

While money supposedly corrupts scientists to disagree with DACC, somehow the vastly larger sums coming from governments, environmental groups, and businesses involved in renewable resources, what might be referred to as "Big Green," are just fine. While $560 million over seven years may seem large, the US government gave just about the same amount ($535 million) to a single company, Solyndra, to encourage the development of renewable resources, money which it lost when the company went bankrupt. The real money is with Big Green, not Big Oil.

Governments spend hundreds of billions of dollars per year on Climate Change and renewable resources to combat Climate Change. Over the years this amounts to trillions of dollars, far beyond what "Big Oil" spends, particularly since many oil companies have a lot invested in alternative energy sources. There are the countless environmental groups whose fundraising is tied to the threat of Climate Change. This money supports much of the research and as we have seen can threaten the careers of any who might dare raise a question.

A large part of the problem can be seen with the IPCC itself. The panel was created based on the presupposition Climate Change exists and therefore it naturally attracted those who believed it was a problem. How likely is it for the IPCC to reach the conclusion DACC is not a threat and ask to be disbanded? Is it any wonder its reports have consistently over-estimated the problem and have been found to exclude evidence conflicting with their purpose?

Ultimately, it is an adherence to an agenda over truth which is the core problem. Hal Lewis summed the problem up in his previously referenced resignation letter this way,

> *As recently as thirty-five years ago, when I chaired the first APS study of a contentious social/scientific issue, The Reactor Safety Study, though there were zealots aplenty on the outside there was no hint of inordinate pressure on us as physicists… How different it is now… the money flood has become the raison d'être of much physics research, the vital sustenance of much more, and it provides the support for untold numbers of professional jobs… It is of course, the global warming scam, with the (literally) trillions of dollars driving it, that has corrupted so many scientists, and has carried APS before it like a rogue wave… I think it is the money, exactly what Eisenhower warned about a half-century ago. There are indeed trillions of dollars involved, to say nothing of the fame and glory (and frequent trips to exotic islands) that go with being a member of the club. When Penn State absolved Mike Mann of wrongdoing, and the University of East Anglia did the same for Phil Jones, they cannot have been unaware of the financial penalty for doing otherwise.*

If we are to believe Big Oil is biased and cannot be trusted because of the millions they spend, what does it say about Big Green and the trillions they spend? What does it say about the renewable resources sector, most of whom could not survive without government support driven by DACC? The financial institutions would benefit from the $2 trillion carbon market and then there is the hundreds of millions of dollars in proposed aid to the developing countries. If the issue is money, it is clear that the money is with Big Green.

Again, these are not reasons to reject DACC, which should be done based only on the evidence, a small part of which I have presented. They are at best reasons to be skeptical of both sides and look at the pros and cons and come to your own conclusions about how to put all the evidence together. Some argue it is the skeptics who are not being scientific, who are being selective in their choice of evidence, etc. Perhaps, but perhaps not.

SUMMARY

DACC rests on the recent increase in temperature, combined with increasing CO_2 and sees the two as linked. Skeptics point out the recent increase started well before human induced CO_2 levels could possibly have had an effect. Skeptics also look at a much longer timespan and see nothing unusual in the recent increase, while also pointing to evidence that the effects of CO_2 are far more complex and not yet well understood. DACC rests on project-ing the recent increase into the future, projections that have been consistently wrong. Skeptics say we do not yet have enough of an understanding to predict the climate. DACC sees only a steady increase in temperature unless CO_2 levels are reduced. Skeptics see the climate continuing to cycle though warming and cooling trends just as it always has, even in the 20th century, and now with the Hiatus.

Finally, I would point out the following known problems with DACC theory as a summary. It is the Hockey Stick that tried to erroneously wipe out the Medieval Warm Period and Little Ice Age, despite the historical record to the contrary. It is the Hockey Stick that is based on a questionable switch from tree ring data to temperature data. It is DACC theory that is inconsistent with the temperature data from ice cores prior to the Little Ice Age. It is DACC theory ignoring the fact Greenland is much colder now than 1000 years ago. It is DACC that changes the language to meet its goals. It is DACC supporters that have for decades been falsely predicting dire effect of cooling then later warming, predictions that have consistently failed. It is DACC supporters who have consistently focused only on the problems of warming while ignoring possible benefits. It is DACC supporters who have consistently downplayed the cost of their proposed solutions which have already caused suffering to millions. It is supporters of DACC who falsely and unscientifically claim there is a consensus in order to try and shut down debate. It is DACC supporters whose models have consistently failed in the short term and yet demand we accept the prediction 50 to 100 year out where they are effectively untestable, and thus unscientific. It is DACC scientists, who try to hide their methods and data so they cannot be tested, and who talk of having too many years invested to let people find problems with their work. It is DACC scientists who talk of tricks and hiding data to get the results they desire. It is DACC supporters who irrationally use ad hominem attacks on those who question their agenda. It is DACC supporters who try to intimidate those who disagree. And it the midst of all this, the climate did not play along warming when they were predicting cooling, and then it stopped warming, contrary to all their models and predictions. And despite all of this, the IPCC makes claims of 95% confidence.

Thus, the question: how many times must they be wrong, before we begin to suspect they may not be right?

FURTHER READING

There are a lot of good sources for questioning DACC, far too many to list here, but here are just four places to start:

Books:

Climate Change: The Facts, edited by Alan Moran
A Disgrace to the Profession, Compiled and Edited by Mark Steyn

Websites:

wattsupwiththat.com
drroyspencer.com

TOPICAL LINE DRIVES
Straight to the Point in under 44 Pages

All Topical Line Drives volumes are priced at $4.99 print and 99¢ in all ebook formats.

Available
The Authorship of Hebrews: The Case for Paul   David Alan Black
What Protestants Need to Know about Roman Catholics   Robert LaRochelle
What Roman Catholics Need to Know about Protestants   Robert LaRochelle
Forgiveness: Finding Freedom from Your Past   Harvey Brown, Jr.
Process Theology: Embracing Adventure with God   Bruce Epperly
Holistic Spirituality: Life Transforming Wisdom from the Letter of James
    Bruce Epperly
To Date or Not to Date: What the Bible Says about Pre-Marital Relationships
    D. Kevin Brown
The Eucharist: Encounters with Jesus at the Table   Robert D. Cornwall
The Authority of Scripture in a Postmodern Age: Some Help from Karl Barth
    Robert D. Cornwall
Rendering unto Caesar   Chris Surber
The Caregiver's Beattitudes   Robert Martin
What is Wrong with Social Justice   Elgin Hushbeck, Jr.
I'm Right and You're Wrong   Steve Kindle
Words of Woe: Alternative Lectionary Texts   Robert D. Cornwall
Stewardship: God's Way of Recreating the World   Steve Kindle
Those Footnotes in Your New Testament   Thomas W. Hudgins
Jonah: When God Changes   Bruce G. Epperly
Ruth & Esther: Women of Agency and Adventure   Bruce G. Epperly
Constructing Your Testimony   Doris Horton Murdoch

Forthcoming
God the Creator: The Variety of Christian Views on Origins   Henry Neufeld

(The titles of planned volumes may change before release.)

Generous Quantity Discounts Available
Dealer Inquiries Welcome
Energion Publications — P.O. Box 841
Gonzalez, FL 32560
Website: http://energionpubs.com
Phone: (850) 525-3916

ALSO FROM ENERGION PUBLICATIONS

Moffett-Moore succeeds in engaging the Christian tradition about Creation in a meaningful dialogue with our contemporary understandings of the natural world.

Herold Weiss, PhD
author of *Creation in Scripture* and Professor Emeritus of New Testament, St. Mary's College, Notre Dame

ALSO BY ELGIN HUSHBECK, JR.

Every citizen who seeks to be not merely informed but also inspired ought to read it.

W. B. Allen
Emeritus Professor of Political Philosophy
Michigan State University